Introdι
Cleanr

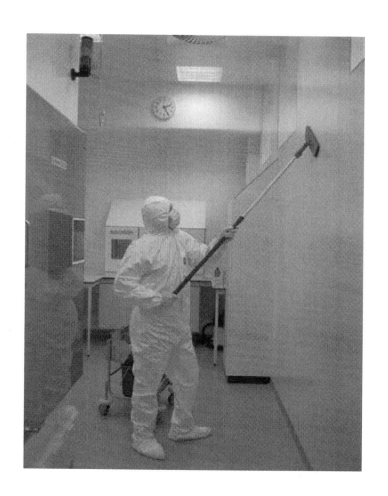

Dr. Tim Sandle

Published by Microbiology Solutions, St Albans, UK

Further information is available from:

www.pharmamicroresources.com

About the author

Dr. Tim Sandle

Dr. Sandle has over twenty-five years experience of microbiological research and biopharmaceutical processing. This includes experience of designing, validating and operating a range of microbiological tests including sterility testing, bacterial endotoxin testing, bioburden and microbial enumeration, environmental monitoring, particle counting and water testing. In addition, Dr. Sandle is experienced in pharmaceutical microbiological risk assessment and investigation.

Dr. Sandle is a tutor with the School of Pharmacy and Pharmaceutical Sciences, University of Manchester for the university's pharmaceutical microbiology MSc course. In addition, Dr. Sandle serves on several national and international committees relating to pharmaceutical microbiology and cleanroom contamination control (including the ISO cleanroom standards). He is a committee member of the Pharmaceutical Microbiology Interest Group (Pharmig); and is a member of several editorials boards for scientific journals. He is also Editor in Chief for SOJ Microbiology & Infectious Diseases.

Preface

This guide has been developed in order to provide a short and informative guide to cleanrooms. Cleanrooms provide critical environments for both sterile and non-sterile pharmaceutical manufacturing.

The guide is divided into different sections. The first section looks at cleanrooms; different grades of cleanrooms and the important aspects of physical control. The second part looks at cleanroom classification. The third part discusses contamination control and environmental monitoring. The fourth part lists the important cleanroom parameters required by the regulatory standards.

Photograph: Technician working in a clean air device

Cleanrooms are highly controlled environments where the air quality is monitored to ensure the extreme standards of cleanliness required for the manufacture of pharmaceutical, electronic and healthcare goods. These stringent standards usually require high fresh air rates, extensive filtering, temperature and humidity control - all of which results in increased energy usage. Protection from uncontrolled ingress of external ambient air is achieved by creating a pressure differential between the cleanroom and its surroundings.

The guide was originally written in 2012. This edition was produced in 2016, to reflect the changes to the international cleanroom standard ISO 14644 (Parts 1 and 2).

Guide to Cleanrooms

Introduction

Contamination control is the primary consideration in cleanroom design; however the relationships between contamination control and airflow are not well understood. Contaminants such as particles or microbes are primarily introduced to cleanrooms by people although processes in cleanrooms may also introduce contamination. During periods of inactivity or when people are not present, it is possible to reduce airflow and maintain cleanliness conditions. To design the cleanroom, the following factors must be accounted for:

- Minimize clean space,
- Correct cleanliness level,
- Optimal air change rate,
- Consider use of mini-environments,
- Optimize ceiling coverage,
- Consider cleanroom protocol and cleanliness class,
- Minimize pressure drop (air flow resistance),
- Location of large air handlers – close to end use,
- Adequate sizing and minimize length of ductwork,
- Provide adequate space for low pressure drop air flow,
- Low face velocity,
- Use of variable speed fans,
- Optimizing pressurization,
- Consider air flow reduction when unoccupied,
- Efficient components,
- Face velocity,
- Fan design,
- Motor efficiency,
- HEPA filters differential pressures (sometimes expressed as ΔP),
- Fan-filter efficiency,
- Electrical systems that power air systems.

The performance of a cleanroom is defined by a set of complex interactions between the airflow, sources of contamination and heat, position of vents, exhausts and any objects occupying the space. Consequently changes to any of these elements will potentially affect the operation of the cleanroom and could invalidate aspects of the room design.

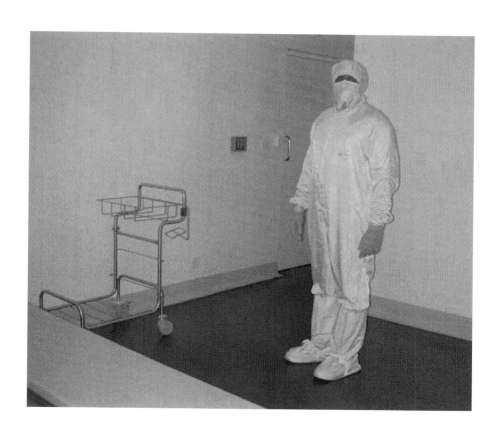

Photograph: Operator in a changing room

Part A: Introduction to cleanrooms

What are cleanrooms?

Cleanrooms and zones are typically classified according to their use (the main activity within each room or zone) and confirmed by the cleanliness of the air by the measurement of particles. Cleanrooms are used in several industries including the manufacture of pharmaceuticals and in the electronics industry. For pharmaceutical cleanrooms, air cleanliness is either based on EU GMP guidance for aseptically filled products (as per Annex 1) and the EU GMP alphabetic notations are adopted; or by using the International Standard ISO14644, where numerical classes are adopted. The U.S. Food and Drug Administration (FDA) uses ISO classes, whereas the World Health Organization (WHO) adopts the EU GMP notation.

The cleanliness of the air is controlled by an HVAC system (Heating, Ventilation and Air-Conditioning).

The key aspect is that the level of cleanliness is controlled.

A more specialised meaning is:

"A room with control of particulates and set environmental parameters. Construction and use of the room is in a manner to minimise the generation and retention of particles. The classification is set by the cleanliness of the air" (as defined in ISO 14644-1).

By prescribing a grade or a class to a clean room, the areas are then regarded as controlled environments. A controlled environment is:

"Any area in an aseptic process system for which airborne particulate and microorganism levels are controlled to specific levels to the activities conducted within that environment" (as used in the Institute of Validation Technology Dictionary).

For EU GMP the typical room uses and associated grades are:

Grade	Room Use
A	Aseptic preparation and filling (critical zones under unidirectional flow)
B	A room containing a Grade A zone (the background environment for filling) and the area demarcated as the 'Aseptic Filling Suite' (including final stage changing rooms)
C	Preparation of solutions to be filtered and production processing; component handling.
D	Handling of components after washing; plasma stripping
U*	Freezers, computer conduits, store rooms, electrical cupboards, other rooms not in use etc.

* U = unclassified. Unclassified areas are not monitored.

Thus Grade A is the highest grade (that is the 'cleanest') and Grade D the lowest (that is the least 'clean'). ISO 14644 equivalents are detailed below.

ISO Class number (N)	Maximum allowable concentrations (particles/m³) for particles equal to and greater than the considered sizes, shown below[a]					
	0,1 μm	0,2 μm	0,3 μm	0,5 μm	1 μm	5 μm
1	10^b	d	d	d	d	e
2	100	24^b	10^b	d	d	e
3	1 000	237	102	35^b	d	e
4	10 000	2 370	1 020	352	83^b	e
5	100 000	23 700	10 200	3 520	832	d, e, f
6	1 000 000	237 000	102 000	35 200	8 320	293
7	c	c	c	352 000	83 200	2 930
8	c	c	c	3 520 000	832 000	29 300
9g	c	c	c	35 200 000	8 320 000	293 000

With the 2015 update to the ISO 14644 standard (Part 1), there is no longer any > 5.0 micron particle concentration for ISO 5. However, those who are required to measure particles at this size need to add it to the classification process. This is explained later.

Another type of cleanzone is an Isolator. Isolators are superior to cleanrooms in that the contamination risk is reduced through the construction of a barrier between the critical area (sometimes called the 'micro-environment') and the outside environment. Isolators are used for sterility testing, aseptic filing and other applications where a clean environment is required. It is important that any possibility of contamination is avoided so that a 'false positive' does not occur.

What are clean air devices?

Within cleanrooms are various clean air devices. The terminology of ISO 14644-7, Cleanrooms and associate controlled environments - Part 7, uses the term 'Separative Devices' to collectively describe clean air hoods, gloveboxes, isolators and minienvironments. These devices include laminar airflows (more commonly described as Unidirectional Airflow (UDAF) Devices in the context of pharmaceutical manufacturing given that 'true' laminarity cannot be easily demonstrated), Biosafety Cabinets and Isolators. Such devices normally operate at EU GMP Grade A / ISO Class 5. The term 'cabinet' is used more widely within Europe and the term 'hood' used more widely in the USA.

Whereas most cleanrooms operate with a turbulent airflow, clean air devices are designed to minimize turbulence which creates dust and dirt collection pockets by operating with the air blowing in one direction (unidirecitonal), where the design feature is to move air away from the critical activity to ensure that any contamination is blown away to a less critical area.

The photograph shows a pharmaceutical process isolator.

With UDAF devices these are either constructed with horizontal flow or vertical flow. Specially designed UDAFs are biosafety cabinets. These are 'self-contained' enclosures which provide protection for personnel, environment and/ or products in work with hazardous microorganisms. The cabinets provide protection by creating an air barrier at the work opening and by high efficiency particulate air (HEPA) filtration of exhaust air. Class I cabinets protect the operation or the product from personnel contamination, whereas Class II cabinets protect personnel, environment and products.

For some UDAF devices, gloves are fitted in order to restrict the number of personnel interventions. Such devices are described as Restrict Access Barrier Systems (RABS). These stand partway between a conventional UDAF and an isolator.

Another special type of cabinet is the powder containment cabinet. These are compact containment cabinets with inward airflow and HEPA filtration that provide protection for operators and the environment from powders generated by processes such as compounding of pharmaceuticals.

Another type of clean air device is an Isolator. Isolators are superior to cleanrooms in that the contamination risk is reduced through the construction of a barrier between the critical area (sometimes called the 'micro-environment') and the outside environment. Isolators are used for sterility testing, aseptic filing and other applications where a clean environment is required. It is important that any possibility of contamination is avoided so that a 'false positive' does not occur.

A variation of an isolator is a glovebox. A glovebox is an enclosure, fitted with sealed gloves, that allows external manual manipulations in controlled or hazardous environments.

What are pass-through or transfer hatches?

Many cleanrooms contain pass-through hatches. These are hatches with double doors that protect critical environments while allowing transfer or materials to or from adjoining rooms. They are typically installed within the walls of cleanrooms. The hatches allow materials to be transferred with minimal loss of room pressure and without the need for personnel movement between rooms.

What are airlocks?

An airlock is an airtight room which adjoins two cleanrooms. The airlock acts as a buffer zone between two independent areas of unequal pressure. A pressure differential of ≥ 15 Pa is typically maintained between the inner room and the air lock; and between the air lock and the external area (see later for information relating to pressure differentials).

What is contamination?

Cleanrooms are designed to minimise and to control contamination. There are many sources of contamination. The atmosphere contains dust, microorganisms, condensates, and gases. Manufacturing processes will also produce a range of contaminants. Wherever there is a process which grinds, corrodes, fumes, heats, sprays, turns, etc., particles and fumes are emitted and will contaminate the surroundings.

People, in clean environments, are the greatest contributors to contamination emitting body vapours, dead skin, micro-organisms, skin oils, and so on. The average person sheds 1,000,000,000 skin cells per day, of which 10% have micro-organisms on them. This demonstrates the importance of wearing cleanroom clothing and wearing this clothing correctly.

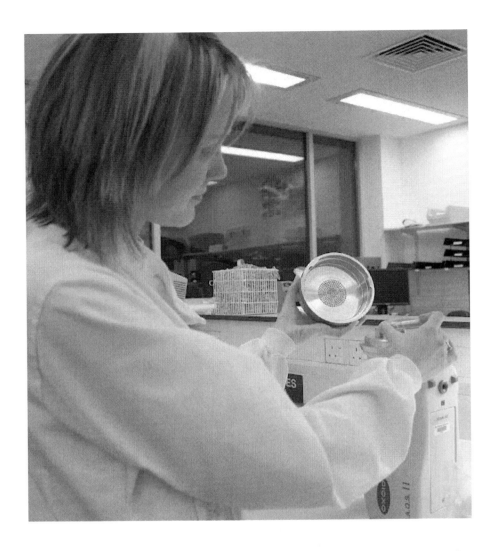

Photograph: Operator preparing an active air sampler for environmental monitoring

Most cleanroom micro-organisms are in the air. If they settle on a dry surface they are unlikely to survive and ideally any contamination is removed from the room. The biggest concern is water, which is both a vector and a growth source for micro-organisms.

Second to people, the key contamination source is water. This is an important issue for water is the main ingredient in many products, and it is used widely throughout the main process areas.

What is contamination control?

Contamination control is critical to all aspects of pharmaceutical manufacturing. Practices are put in place to ensure that the air is of the correct standard; that opportunities for contamination are not present (like water puddles on the floor); and that contamination carried on people is minimised.

Other ways by which contamination is controlled are:

- The air entering a cleanroom from outside is filtered to exclude dust, and the air inside is constantly re-circulated through HEPA filters. This is controlled through a HVAC (Heating, Ventilation and Air Conditioning) system. The most important part of this is with air-filtration through a HEPA (High Efficiency Particulate Air) filters.
- Staff enter and leave through airlocks and wear protective clothing such as hats, face masks, gloves, boots and cover-alls.
- Equipment inside the cleanroom is designed to generate minimal air contamination. There are even specialised mops and buckets. Cleanroom furniture is also designed to produce a low amount of particles and to be easy to clean.
- Common materials such as paper, pencils, and fabrics made from natural fibres are excluded from the Aseptic Filling Suite.
- Some cleanrooms are kept at a higher air pressure so that if there are any leaks, air leaks out of the chamber instead of unfiltered air coming in.
- Cleanroom HVAC systems also control the humidity to low levels, such that extra precautions are necessary to prevent electrostatic discharges.

Contamination control also requires personnel to practice aseptic techniques; wear specially designed clothing; to clean the areas to the correct standard; and to behave in ways which will minimise contamination.

Why monitor air quality in cleanrooms?

Air is both a means to ensure that cleanrooms are clean and it can be a source of contamination. Air cannot be avoided for without air we cannot breathe so as long as we require personnel to operate our processes we need an air supply.

Even in clean rural areas air is contaminated with about 108 particles of 0.5µm and greater per m^3, many of these will be microorganisms depending on the nature of the area and the season of the year: so air is a contamination problem. However in the pharmaceutical industry air flow is the answer to many contamination problems (as discussed in relation to the physical monitoring of cleanrooms below).

In order to ensure that cleanrooms are operating correctly, air is monitored through:

- Formal classification of cleanrooms (as defined by air cleanliness, which relates to the number of airborne particles)
- Through physical measurements of HVAC operations
- Through non-viable particle monitoring
- Through viable particle monitoring

There are four principles applying to control of air-borne microorganisms in clean rooms.

- Filtration (through the use of HEPA filters)
- Dilution (to ensure that particles generated in clean rooms, in addition to those which pass the filters, are carried away by diluting the clean area with new "clean" air)
- Directional Air Flow (to ensure that air blows away from critical zones, as particles and microorganisms cannot "swim upstream" against a directional air flow)

- Air Movement (rapid air movement is important for as long as particles and microorganisms stay suspended in the air they are not really a problem, for it is only when they settle out that they become an actual cause of contamination)

Photograph: An operator cleaning within a cleanroom

Different operating conditions

Clean rooms have three different 'states' of use. These are:

- As built;
- Static;
- Dynamic.

As built refers to the condition of a newly built clean room, with the operational qualification having been completed, at the point it is handed over to the user for performance qualification.

For static conditions is the room without personnel present, following 15 – 20 minutes 'clean up time', but with equipment operating normally.

Dynamic conditions (or 'operational') are defined as rooms being used for normal processing activities with personnel present and equipment operating.

Photograph: A technician preparing an isolator for testing

Part B: Cleanroom standards

Prior to 1999, the main cleanroom standard for the classification of cleanrooms was U.S. Federal Standard 209 "Airborne Particulate Cleanliness Classes in Cleanrooms and Clean Zones." The last edition to be published was version E, on June 16, 1988. EU GMP, it should be noted, dealt with the assessment of cleanrooms during operation specifically for the aseptic filling of medicinal products.

As time progressed cleanrooms became more sophisticated and widespread. This led to international agreement that there needed to be a wider range of standards, and for this update to take the form of a single standard to cover classification and testing.

This lead to the formation of ISO 14644. The standard is applicable across several industries: covering healthcare, pharmaceuticals and electronics; and it outlines the approach for the design, classification and operation of cleanrooms.

The first part of the standard to be issued was ISO14644-1. The U.S. General Services Administration (GSA) released a Notice of Cancellation for FED-STD-209E, Airborne Particulate Cleanliness Classes in Cleanrooms and Clean Zones, on November 29, 2001 and FED-STD- 209E was then superseded by ISO 14644-1 and ISO 14644-2.

Part 1 of ISO 14644 determined the method by which a room should be classified, which is by the maximum allowable particles within a fixed volume of air. This was against one of three occupancy states:

- As Built, a completed room with all services connected and functional, but without production equipment or personnel within the facility
- At Rest, a condition where all the services are connected, all the equipment is installed and operating to an agreed manner, but no personnel are present.

- Operational, all equipment is installed and is functioning to an agreed format, and a specified number of personnel are present working to an agreed procedure.

Classification is the process of qualifying the cleanroom environment by the number of particles using a standard method. The end result of the activity is that cleanroom x is assigned ISO class y. Importantly classification is distinct from routine environmental monitoring and distinct from process monitoring, such as the requirement in EU GMP Annex 1 for continuous monitoring of aseptic filling.

The fit between ISO 14644 classes and EU GMP grades has always been inelegant. The complication is primarily because the EU GMP grade is the same for a room at rest and in operation, whereas the ISO class shifts. For example, EU GMP Grade C 'at rest' is equivalent to ISO class 7; whereas the grade C room 'in operation' is equivalent to ISO class 8.

From the issue of Part 1, other parts followed and today there are 12 active parts of the standard. Part 2, which has also been subject to a recent revision, specifies requirements for periodic testing of a cleanroom or clean zone to prove its continued compliance with ISO 14644-1 for the designated classification of airborne particulate cleanliness.

For reference, the full set of ISO 14644 parts is listed below:

- ISO 14644-1:2015 - Part 1: Classification of air cleanliness.
- ISO 14644-2:2015 - Part 2: Specifications for testing and monitoring to prove continued compliance with ISO 14644.
- ISO 14644-3:2005 - Part 3: Test methods.
- ISO 14644-4:2001 - Part 4: Design, construction and start-up.
- ISO 14644-5:2004 - Part 5: Operations.
- ISO 14644-6: 2004 - Vocabulary.
- ISO 14644-7:2004 - Part 7: Separative devices (clean air hoods, gloveboxes, isolators and mini-environments).

- ISO 14644-8:2013 - Part 8: Classification of air cleanliness by chemical concentration (ACC).
- ISO 14644-9:2012 - Part 9: Classification of surface cleanliness by particle concentration.
- ISO 14644-10:2013 - Part 10: Classification of surface cleanliness by chemical concentration.
- No part 11 in draft.
- ISO 14644-12:draft - Part 12: Classification of air cleanliness by nanoscale particle concentration .
- ISO 14644-13:draft - Part 13: Cleaning of surfaces to achieve defined levels of cleanliness in terms of particle and chemical classifications.
- ISO 14644-14:draft - Part 14: Assessment of suitability for use of equipment by airborne particle concentration.

Importantly ISO 14644 is NOT a GMP standard. Parts of the standard have been adopted by GMP systems, such as the reference to ISO classes in the FDA Guidance on Aseptic Processing and the requirement to use the standard to classify cleanrooms in Annex 1 of EU GMP. Not all of the parts of the standard are applicable to GMP environments, for example, part 12 is intended for the nanotechnology industry.

In December 2015, Parts 1and 2 of ISO 14644 were revised. The more substantial changes relate to Part 1. As part of the change process, the title of the second part of the standard standards was altered to: "Specifications for testing and monitoring to prove continued compliance by ACP" (with ACP representing 'airborne particulate contamination').

Part C: Key cleanroom parameters

In order to ensure that cleanrooms and their HVAC systems are functioning correctly, they are classified at different intervals. Classification of cleanrooms is confirmed in the dynamic state by taking non-viable particulate readings at a defined number of locations for 5.0μm and 0.5μm size particles.

Once a room has been assigned a classification, certain environmental parameters (physical and microbiological) are to be met on a routine basis. For viable monitoring it is normal for the microbiologist to set action levels based on an historical analysis of data.

The frequency of the assessment of these other parameters should be assessed based on a risk management approach. This approach should consider the room use and the risk to the product. Factors to consider may include room activities; exposure risk; room temperature; process stage; duration of process activities; water exposure and so on.

The recommended emphasis is upon environmental control rather than simply environmental monitoring.

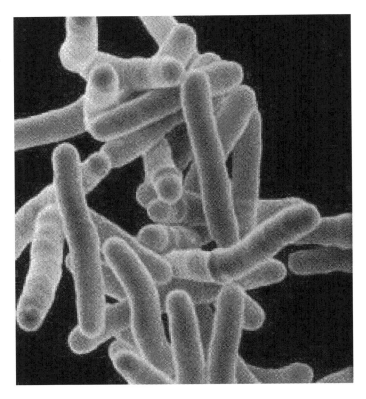

Photograph: Electron micrograph of a Bacillus spp. (bacteria), representative of cleanroom contamination.

HVAC operation

There are a number of physical parameters which require examination on a regular basis. These parameters generally relate to the operation of HVAC systems and the associated air-handling units. Air handler, or air handling unit (AHU) relates to the blower, heating and cooling elements, filter racks or chamber, dampers, humidifier, and other central equipment in direct contact with the airflow.

The picture above shows typical AHU components:

1 - Supply duct
2 - Fan compartment
3 - Flexible connection
4 - Heating and/or cooling coil
5 - Filter compartment
6 - Return and fresh air duct

The key areas of HVAC operation essential for contamination control are:

Air-patterns and air-movement

Airflows, for critical activities, need to be studied in order to show that air turbulence does not interfere with critical processes by mapping smoke patterns. There are two types of cleanroom: turbulent flow or unidirectional flow, depending upon the required application. Unidirectional airflow areas are used for higher cleanliness states (such as aseptic filling) and they use far greater quantities of air than turbulent flow areas.

Airflows

Grade A zones (unidirectional airflow devices in Grade B rooms) have a requirement for controlled air velocity and unidirectional air flow (either horizontal or vertical). These are monitored using an anemometer. The air velocity is designed to be sufficient to remove any relatively large particles before they settle onto surfaces.

The picture above shows a simple unidirectional airflow concept.

Air changes

Each clean room grade has a set number of air changes per hour. A typical air conditioned office will have something between two and ten air changes per hour in order to give a level of comfort. The number of required air changes in a cleanroom is typically much higher. Air changes are provided in order to dilute any particles present to an acceptable concentration (thus air change is a way of expressing the level of air dilution which is occurring).

Clean up times

Connected to air changes is the time taken for a clean area to return to the static condition, appropriate to its grade, in terms of particulates.

Positive Pressure

Connected to the measurement of air flow is positive pressure. In order to maintain air quality in a clean room the pressure of a given room must be greater relative to a room of a lower grade. This is to ensure that air does not pass from " dirtier" adjacent areas into the higher grade clean room (this can also be observed by smoke studies). Generally this is 15-20 Pascals, although some areas of the same grade will also have differential pressure requirements due to specific activities. The most commonly encountered problems relate to situations when cleanroom doors are opened and here it can be difficult to maintain pressures.

HEPA filters

HEPA (High Efficiency Particulate Air) filters are used in to provide clean air to the cleanroom. HEPA filters are replaceable, extended-media, dry-type filters in rigid frames with set particle collection efficiencies. The filters are designed to control the number of particles entering a clean area by filtration. In Grade A zones HEPA filters also function to straighten the airflow as part of the unidirectional flow. In order to measure the effectiveness of the filters they are checked for leaks. Leakage is assessed by challenging the filters with a particle generating substance and measuring the efficiency of the filter.

HEPA filters function through a combination of three important aspects. First, there are one or more outer filters that work like sieves to stop the larger particles of dirt, dust, and hair. Inside those filters, there is a concertina - a mat of very dense fibres - which traps smaller particles. The inner part of the HEPA filter uses three different mechanisms to catch particles as they pass through in the moving airstream. At high air speeds, some particles are caught and trapped as they smash directly into the fibres, while others are caught by the fibres as the air moves past. At lower air speeds, particles tend to wander about more randomly through the filter (via Brownian motion) and may stick to the fibres as they do so. Together, these three mechanisms allow HEPA filters to catch particles that are both larger and smaller than a certain target size.

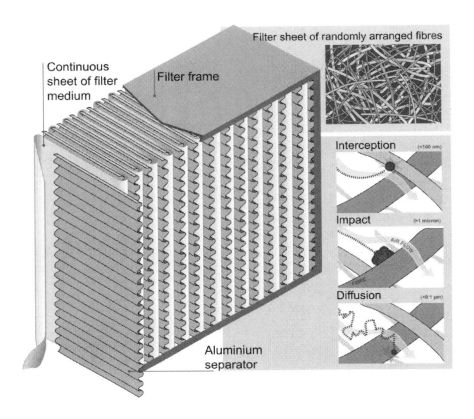

The image shows how a HEPA filter captures and traps contamination.

There are different grades of HEPA filters based on their 'efficiency ratings'. One of the most commonly used HEPA filter is the H14 filter, which is designed to remove 99.997% of particles from the air. HEPA Filters are protected from blockage by pre-filters which remove up to about 90% of particles from air.

To use an example, if air contains about 3×10^8 particles per m³, and there is one pre-filter and one HEPA Filter:

- Pre-filter leaves about 3×10^7 per m³ as a challenge to the HEPA filter

- The terminal HEPA Filter leaves about 10^3 per m³.

- In EU GMP this is within the limits for Grade A and B "at rest" (Annex 1.4)

In Grade A zones HEPA filters also function to straighten the airflow as part of the unidirectional flow. In order to measure the effectiveness of the filters they are checked for leaks. Leakage is assessed by challenging the filters with a particle generating substance and measuring the efficiency of the filter.

Other factors

For certain cleanrooms temperature, humidity and lighting require control, either because of a process step or as a means to minimise contamination.

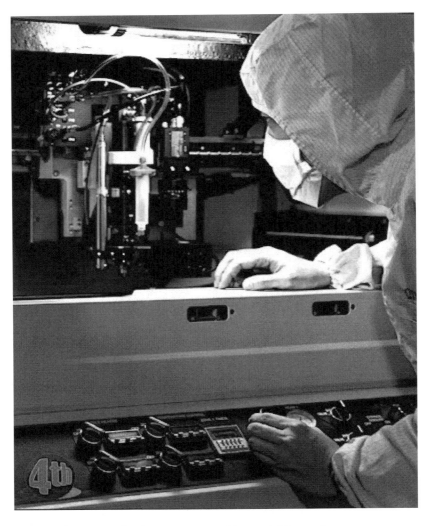

Photograph: Operator working within a cleanroom

Part D: Contamination Control

Contamination

Cleanrooms are designed to minimise and to control contamination. There are many sources of contamination. The atmosphere contains dust, microorganisms, condensates, and gases. Manufacturing processes will also produce a range of contaminants. Wherever there is a process which grinds, corrodes, fumes, heats, sprays, turns, etc., particles and fumes are emitted and will contaminate the surroundings.

People, in clean environments, are the greatest contributors to contamination emitting body vapours, dead skin, micro-organisms, skin oils, and so on. The average person sheds 1,000,000,000 skin cells per day, of which approximately 10% have micro-organisms on them. This demonstrates the importance of wearing cleanroom clothing and wearing this clothing correctly.

Most cleanroom micro-organisms are in the air. If they settle on a dry surface they are unlikely to survive and ideally any contamination is removed from the room. The biggest concern is water, which is both a vector and a growth source for micro-organisms.

Second to people, the key contamination source is water. This is an important issue for water is the main ingredient in many products, and it is used widely throughout the main process areas.

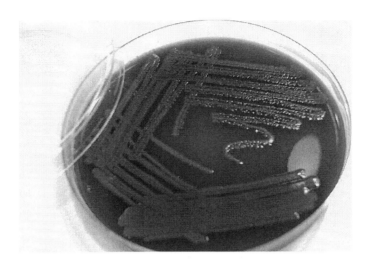

Photograph: microbiological agar plate for the isolation of a contaminant back in the microbiology laboratory.

Contamination control

Contamination control is critical to all aspects of pharmaceutical manufacturing. Practices are put in place to ensure that the air is of the correct standard; that opportunities for contamination are not present (like water puddles on the floor); and that contamination carried on people is minimised.

Other ways by which contamination is controlled are:

- The air entering a cleanroom from outside is filtered to exclude dust, and the air inside is constantly re-circulated through HEPA filters. This is controlled through a HVAC (Heating, Ventilation and Air Conditioning) system. The most important part of this is with air-filtration through a HEPA (High Efficiency Particulate Air) filters.
- Staff enter and leave through airlocks and wear protective clothing such as hats, face masks, gloves, boots and cover-alls.
- Equipment inside the cleanroom is designed to generate minimal air contamination. There are even specialised mops and buckets. Cleanroom furniture is also designed to produce a low amount of particles and to be easy to clean.
- Common materials such as paper, pencils, and fabrics made from natural fibres are excluded from the Aseptic Filling Suite.
- Some cleanrooms are kept at a higher air pressure so that if there are any leaks, air leaks out of the chamber instead of unfiltered air coming in.
- Cleanroom HVAC systems also control the humidity to low levels, such that extra precautions are necessary to prevent electrostatic discharges.

Contamination control also requires personnel to practice aseptic techniques; wear specially designed clothing; to clean the areas to the correct standard; and to behave in ways which will minimise contamination.

Contamination incidents

There are several types of contamination incident that can occur within cleanrooms. These tend to occur because:

a) Water is left on floors, which allows micro-organisms to grow;
b) Equipment is used wet;
c) Equipment which is clean has not been properly segregated from equipment which is wet;
d) Cleaning and disinfection is not undertaken to defined frequencies;
e) Air-lock doors do not open or close properly, allowing contaminated area to move between areas;
f) Personnel are not dressed correctly and allow contamination to spread.

Environmental monitoring

Environmental monitoring is a programme which examines the numbers and occurrences of viable micro-organisms and 'non-viable' particles (that is particles in the area other than micro-organisms like dust or skin cells). Ideally, environmental monitoring is targeted to those areas of the production process where the risk cannot be adequately controlled. It thereby, through trend analysis, provides an indication if the cleanroom is moving out-of-control.

Non-viable monitoring is for air-borne particle counts. These are the same sizes of particles required for the classification: 0.5 and 5.0 µm. This is undertaken using an optical particle counter. Particle counters are used to determine the air quality by counting and sizing the number of particles in the air.

What are particles?

'Particle' in the context of a cleanroom is a general term for subvisible matter. Airborne Particles, refers to particles suspended in air. Air contains a variety of different particles of a range of different sizes. These are particles of dust, dirt, skin, microorganisms and so on. As discussed above, the function of cleanrooms is to reduce the number of airborne particles (for example, an office building air contains from 500,000 to 1,000,000 particles (0.5 microns or larger) per cubic foot of air. In contrast, an ISO Class 5 / EU GMP Grade A cleanroom is designed not to allow more than 100 particles (0.5 microns or larger) per cubic foot of air.

With cleanrooms the regulatory standards, which are discussed below, focus on two sizes of particles which are selected due to the potential risk that they pose. These are:

- 0.5μm size particles, which are close in size to many microorganisms;
- 5.0 μm size particles, which are close in size to skin flakes, on which many microorganisms are bound to.

With European GMP, there is concern with both types of particle size. With the FDA, the primary focus is upon the 0.5μm size.

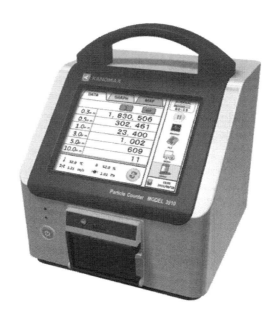

Photograph: Particle counter

What is the unit of measurement for particles?

The unit of measurement for particles is the micrometer (or 'micron'). This is symbolised as μm. The micron is a unit of length equal to one millionth (10^{-6}) of a metre.

What are the Sources of Particles in Cleanrooms?

Particles are generated from a variety of sources. These can include:

- Facilities, such as: walls, floors and ceilings; paint and coatings; construction material; air conditioning debris; room air and vapours; spills and leaks.
- People, including: skin flakes and oil; cosmetics and perfume; spittle; clothing debris (lint, fibres etc.); hair.
- Equipment generated, including: friction and wear particles, lubricants and emissions, vibrations
- Cleaning equipment, like: brooms, mops and dusters; cleaning chemicals
- Fluids, arising from spillages
- Particulates floating in air, primarily: bacteria, fungi, organic material and moisture
- Compressed gasses
- Product generated

Within cleanrooms the primary concern is with those particles which are microorganisms or likely to be carrying microorganisms, such as skin flakes. The major source, and hence the primary risk, is from people. The risk can be increased through physical behaviour like fast motion and horseplay or from physiological concerns like room temperature, humidity or from psychological concerns like claustrophobia, odours and workplace attitude. In general, people produce contamination via:

- Body Regenerative Processes: skin flakes, oils, perspiration and hair.
- Behaviour: rate of movement, sneezing and coughing.

- Attitude: work habits and communication between personnel.

A degree of protection is provided through cleanroom clothing. Cleanroom gowns are manufactured from special materials which are designed to minimise the amount of contamination which can be shed from the skin, provided that the gown is not worn of an excessive time and that the temperature is not too high. Special apparel includes non-shedding gowns or coveralls, head covers, face masks, gloves, footwear or shoe covers.

What are particle counters?

Particle counting is performed using a variety of optical particle counters (aerosols passed through a focused light source, where the scattered light is converted into electrical pulses which allow the counting of particles). These are designed to detect the number of particles of a given size from a given volume of air. The types of counters used are detailed in Microbiology method SOPs. Some particle counters maybe connected to a Facility Monitoring System (FMS).

A particle counter is a device which draws air in using a pump at a controlled flow rate. The air is passed into a sensor area and through a light beam created by a laser diode. The amount of light reflected from each particle is measured electronically (as an electronic pulse). The larger the particle, then the larger the amount of reflected light (the greater the height of the light pulse). This allows the particle counter to 'count' the number of particles in a given volume of air (as the number of light pulses) and to assess the size of the particles counted.

Different particle counters have different **flow rates**. The flow rate is the rate at which air is drawn into a particle counter, and thus the time taken for the counter to measure a fixed volume of air. The long standing flow rate has been 1.0 cubic feet per minute (equivalent to 28.3 litres per minute). This flow rate is the baseline for cleanroom certification.

What is particle loss?

Particle loss is minimised by the use of specialised tubing (such as Bev-a-line or Tygon tubing) which is designed to prevent particles from adhering to the tubing surface. Thus the quality of the material used for particle counter tubing is important. In general, there are three types of tubing which may be considered:

a) Bev-A-Line tubing or Tygon tubing

   Bev-A-Line or Tygon tubing tubing is a co-extruded tubing consisting of a PVC exterior and a Hytrel interior. Its suitability as a tubing for particle counters relates to the smoothness of the interior wall.

b) Stainless steel

   Stainless steel tubing is suitable for situations where particles in a hot air-stream require measurement (such as a dehydrogenation tunnel). The disadvantage of the tubing is its lack of flexibility.

c) Polyurethane

Particle loss can also occur due to tubing diameter. The recommended internal diameter of tubing for particle counters, by particle counter manufacturers, is ¼". This may vary depending upon the flow rate of the counter (Pollen, a, p2). The tubing used at BPL has a diameter of 10-15mm (refer to approved purchase register). This ensures that the tubing has the correct Reynolds number.

Particle counter tubing lengths must be kept as short as possible. This is particularly important for avoiding particle loss for particles of a size of >1.0 micron. Research suggests that there is a 20% loss of 5.0 micron particle counts for tubing lengths of >3 metres (approximately 10 feet). Particle counter tubing should not exceed 3 metres in length between the sampling head probe and the particle counter. This is to ensure the transportation and delivery of larger particles (such as 5.0 µm) and to avoid 'drop-out'. Tubing should also be as straight as possible. Radial bends will result in the loss of particles (Pollen, a, p3)

Particle counters tubing must be changed at regular intervals (such as three-monthly). All particle counter tubing will, over time, accumulate particles, particularly where particle counters are used for continuous monitoring. A phenomenon which can arise is the sudden release of particles (previously suspended on tubing walls) which may lead to an unusually high count or series of counts.

What different counting modes do particle counters have?

Particle counters can be set for one of two counting modes:

i) **Cumulative count**: where the counter is set to count the number of particles for the selected size and greater. For example, if a counter is set to count 0.5 μm particles, it will count all particles at the 0.5 μm and greater (such as 0.5, 0.7, 1.0, 5.0 and 10.0, depending upon the number of available channels on the counter). For cleanroom classification and for particle monitoring for EU GMP, the cumulative mode must always be used.

ii) **Differential mode**: where the counter is set to only count the number of particles of the selected size. For example, if a counter is set to count 0.5 μm particles, it will only count particles of the 0.5 μm size.

What is microbiological environmental monitoring?

Environmental monitoring is a programme which examines the numbers and occurrences of viable micro-organisms (as wells as 'non-viable' particles as discussed above). Ideally, environmental monitoring is targeted to those areas of the production process where the risk cannot be adequately controlled. It thereby, through trend analysis, provides an indication if the cleanroom is moving out-of-control.

What is viable monitoring?

Viable monitoring is designed to detect levels of bacteria and fungi present in defined locations /areas during a particular stage in the activity of processing and filling a product. Samples are taken from walls, surfaces, people and the air (each of which represents a potential contamination source). Viable monitoring is designed to detect micro-organisms and answer the questions: how may?; how frequent?; when do they occur?; why do they occur?

Viable monitoring is undertaken using a substance called agar (a jelly-like growth medium) in different sized containers. Sometimes mechanical devices are used to pull in a defined quantity of air (an air-sampler).

The environmental monitoring programme is normally controlled by the Microbiology Department who establish the appropriate frequencies and durations for monitoring based on a risk assessment approach. The sampling plan takes into account the cleanliness level required at each site to be sampled.

What are the methods used for viable monitoring?

Viable microbiological monitoring is normally performed using the following methods:

Table:

Method	Air	Surface	Personnel
1	Active air Sampler (cfu/m³)	Contact Plate (cfu/25cm²)	Finger plate for Hands (cfu/5 fingers) Contact plate for gowns (cfu/25cm²)
2	Settle Plate (cfu/90mm over x time)	Swab (cfu/surface)	

With settle plate exposure, normally the exposure time is per four hours. With the exception of short duration events (such as filtration of product into the filling suite or filling machine set-up) results are expressed as cfu / 4 hours. Where exposure is less than 4 hours ('routine' monitoring uses a time of not less than one hour) results are extrapolated to cfu / 4 hours. For the exceptions mentioned, results are reported as cfu / event. The reason for using cfu / event is that where activities are less than 30 minutes duration there remains a danger that a degree of distortion would occur which would lead to a result which was not commensurate with the risk.

Where cfu = Colony Forming Unit

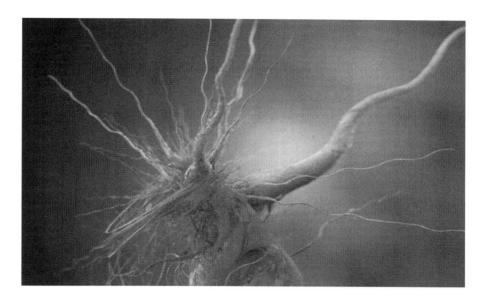

Photograph: representative image of a bacterium

Part E: Critical Cleanroom Parameters

Cleanroom classification

This section of the guide looks at cleanroom classification. Because of the change from the 1999 version of ISO 14644 Parts 1 and 2 in later 2015, the section discusses the changes as well as the essential requirements. This is to assist the reader in understanding the more important changes.

The two main ways by which cleanrooms are classified. This is either to EU GMP or to ISO 14644. There are differences between the two standards whether cleanrooms are operating in the static or dynamic states.

When referring to room grades, the following are equivalent under **'at rest' or static** conditions:

EU GMP	ISO 14644-1
A	5
B	5
C	7
D	8

When referring to room grades, the following are equivalent under **'in operation' or dynamic** conditions:

EU GMP	ISO 14644-1
A	5
B	7
C	8
D	9

Cleanrooms can be classified in one of three occupancy states, as built, static or dynamic. It is more typical for clean rooms to be classified in the **dynamic** state (or 'operational') by taking non-viable particulate readings at a defined number of locations for 5.0μm and 0.5μm size particulates (as defined in a sampling plan) at the following approximate frequencies (as stated in ISO 14644-2):

Grade	Frequency of classification
A	Six-monthly
B	Six-monthly
C	Annually
D	Annually

With frequencies, the general guidance in Part 2 is for annual classification; however, some national standards bodies have elected to have a six-monthly requirement. Six-monthly is quoted because this is the highest standard.

Particle sizes

Cleanroom users can elect to look at one or more particle sizes. With GMP, those who need to meet FDA requirements only will continue to look for particles equal to or greater than 0.5μm. For EU GMP, particles of equal to or greater than 0.5μm and equal to or greater than 5.0μm are a requirement. This falls within what is acceptable by the standard. Here the standard indicates that more than one particle size can be used, provided the next particle size selected is 1.5 times larger than the previous one.

There is an issue with Grade A environments, for particles equal to or greater than 5.0 μm there is no longer any limit set within the standard. This is because of the view that looking for, what would be a low number of particles (less than 20 in a cubic metre of air), is statistically insignificant. However, the standard retains the option for the 5.0 μm particles to be used for classification purposes, but it is up to cleanroom users to select their own limits.

For continuous monitoring, EU GMP inspectors will expect both cut-off particle sizes to continue to be monitored. How classification and batch specific monitoring will fit together might be clarified in the forthcoming update to EU GMP Annex 1.

The ISO 14644-1:2015 standard describes any particle with an equivalent diameter ≥5.0μm as a macro-particle. Where a regulatory agency demands consideration of these particles, the counting and sizing of these macro-particles is expressed using the M descriptor in the format:

ISO M (a; b); c

Where:

1. is the maximum permitted concentration of macro-particles (expressed in particles/m³);
2. is the equivalent diameter of the macro-particles;
3. is the specified measurement method (typically Light Scattering Airborne Particle Counter (LSAPC)).

For example, the Grade A at rest concentration of 20 particles/m³ @ ≥5.0µm would be expressed as:

ISO M (20; ≥5.0 µm); LSAPC

The new M descriptor will be used to define the Annex 1 Grade A and Grade B and by using the existing ISO 14644-1 designation of airborne particle concentration (expressed as *ISO Class number; occupancy state; considered particle size(s)*.

For EU GMP this means:

Grade A

ISO 5; at rest, operational; ≥0.5µm
ISO M (20; ≥5.0µm); at rest, operational; LSAPC

Grade B

ISO 5; at rest; ≥0.5µm and ISO M (29; ≥5.0µm); at rest; LSAPC
ISO 7; operational; ≥0.5µm, ≥5.0µm

Number of particle counter locations

A significant change with the standard is the method for selecting the number (and position) of particle counter locations within a cleanroom. The 1999 approach was that the user calculated the surface area of the cleanroom in square metres. From this the square root was taken and the number generated provided the number of particle counter locations. These are then placed at equidistant intervals within the cleanroom.

With the 2015 revision, the method is based on a look-up table (for cleanrooms of as size up to 1000 square metres - for larger cleanrooms a calculation is required). The table uses a range of cleanroom sizes and provides the number of locations required (if the exact room size is not listed, the user select the next largest room size and pick the appropriate number of locations). These numbers are based on a statistical method called hypergeometric distribution. This is very different to the square root approach, which was based on bionominal distribution. Without going into statistical detail, the former approach assumed that in each location a particle counter was placed, the particles in the cleanroom were normally distributed. In contrast, the revised approach is based on particles not being normally distributed. The new approach allows each location to be treated independently.

For the user, the approach is simpler because no calculations are required. In addition, for rooms with less than 9 particle count locations, the requirement to perform a 95% upper confidence level check has been removed. With the assigned numbers there is an in-built confidence interval of 95%. This means when a cleanroom is monitored there is a 95% level of confidence that 90% of the cleanroom complies.

In general, the new approach leads to an increase in particle count locations compared with the previous standard.

As an example, consider three cleanrooms coded A, B and C (Table 1).

Cleanroom	Room size	1999 version location numbers	2015 version revised no. of locations
A	200 m^2	15	23
B	36 m^2	6	9
C	8 m^2	3	4

Table 1: Cleanroom classification examples.

With cleanroom A, which has a size of 200 square metres, the number of locations was 15. With the revised version, it has risen to 23. You can see similar increases for rooms B and C.

Readers who were familiar with the 1999 version of the standard should note the requirements to calculate the 95% upper confidence limit(s) for 2 to 9 sample locations was removed from the 2015 edition.

Location of particle counters

Once the number of locations has been selected, the room is divided up into sectors and a particle counter placed in each sector. With the previous standard, these sectors were equal in size and a counter placed in approximate centre. With the revised standard, the position where the counter is placed within each sector is determined by the user. The standard allows counters always to be placed at the same point within the sector; randomly placed within the sector; or evenly distributed; or selected by risk.

The risk based approach would be the best one to adopt. A risk based decision could be based on variables like: room layout; equipment type; airflow patterns; position of air supply and return vents; air-change rates; and room activities.

The reason for not selecting the centre of the location relates back to the issue of particle distribution: particles counts no longer assumed to be homogenous within a sector.

Furthermore, additional locations can be added at the discretion of the facility. This might arise from the room-by-room risk assessment.

In assessing the location, using risk assessment the following steps should be followed:

1. Perform a risk assessment to understand, evaluate and document the potential for adverse contamination events.
2. Develop a written monitoring plan.
3. Review and approve the plan.
4. Implement the plan by performing the monitoring.
5. Analyse, trend (where appropriate) and report performance.
6. Document and implement actions or corrective actions required.
7. Perform periodic reviews of the monitoring plan.

Volume of air to be sampled

A further change is with the volume air that requires sampling in each location. The theory behind this is that the volume of air sampled needs to be sufficient to detect at least 20 particles of the largest particle size selected. The revised standard supplies a formula to be used. The outcome is the number of litres that need to be sampled in each location. The standard requires a minimum of 2 litres per location; the application of the formula can result in this being higher. Generally the lower the particle count limit, the greater the volume to be sampled (so a larger volume is taken from an EU GMP Grade B room compared with a Grade C room).

Assessing the results

There is also a revised way for assessing data. This involves:

1. Recording the results for each location.
2. Covert the results to a cubic metre sample per room.
3. This is set out using the formula published in the standard.
4. It should be noted that the formula states "number of particles at each location or average". This is because an option exists to add more than one particle count location per sector. When this occurs the results are averaged and the average used as the number to proceed with the above calculation. Individual results may fall outside of the class, provided that the mean is within.
5. Although assessment is based on an average, each individual result must be within limits. Those out of limits need to be investigated.

Therefore is no longer a grand total. With the previous version of the standard, provided the total was within limits the room would pass. With the revised standard, each individual result must comply.

Particle counter probe locations

There is new advice about the orientation of the particle counter probe. The counter probe must be orientated into the airflow (for unidirectional air) or pointed upwards for turbulent flow air.

Test certification requirements

The revised standard sets out the requirements for test certificates in relation to cleanroom classification. Certificates must state:

- Name and address of the testing organization.
- Date of testing.
- No. and year of the publication of the relevant part of ISO 14644 e.g. ISO 14644: 1 – 2015.
- Location of cleanroom (or clean zone).
- Specific representation of locations e.g. diagram.
- Designation of cleanroom.
- ISO class (plus EU GMP)
- Occupancy.
- Particle count sizes considered.
- Test method used (and any departures or deviations).
- Identification of test instrument and calibration certificate.
- Test results.

Photograph: HEPA filter

Particle count limits

For those who are involved with aseptic processing, and need to draw an EU GMP - ISO 14644 comparison, the following limits are the maximum levels allowed in a clean room, as per EU GMP Guide.

Parameters	Grade A		Grade B		Grade C		Grade D	
Non-viable particulates Static State	Particle size/m^3		Particle size/m^3		Particle size/m^3		Particle size/m^3	
	3,520 at 0.5μm	20 at 5.0μm	3,520 at 0.5μm	29 at 5.0μm	352,000 at 0.5μm	2,900 at 5.0μm	3,520,000 at 0.5μm	29,000 at 5.0μm
Non-viable particulates Dynamic State	Particle size/m^3		Particle size/m^3		Particle size/m^3		Particle size/m^3	
	3,520 at 0.5μm	20 at 5.0μm	352,000 at 0.5μm	2,900 at 5.0μm	3,520,000 at 0.5μm	29,000 at 5.0μm	Not defined *	Not defined *

C/m^3 = count per cubic metre

Airflows

Airflows, for critical activities in relation to aseptic filling, need to be studied in order to show that air turbulence does not interfere with critical processes. All critical rooms and zones within the Aseptic Filling area relating to batch filling will be assessed by Microbiology. Other critical processes may also be monitored.

Specification	Source of specification	Frequency	Method
Visual assessment of risk to process	None – although recommended practice.	To be determined by the user.	Smoke generation and mapping

Note: For air flow movement, air flow must be from a higher grade area to a lower grade area.

Air velocity

Grade A zones (undirectional airflow devices in Grade B rooms) have a requirement for controlled air velocity and unidirectional air flow (either horizontal or vertical). These are monitored using an anemometer. The air velocity is designed to be sufficient to remove any relatively large particles before they settle onto surfaces.

This monitoring is performed routinely and during re-qualification exercises.

Specification	Source of specification	Frequency	Method
0.45 m/s +/- 20% (unidirectional)	EU GMP	Six monthly	Anemometer

The FDA Guide to Aseptic Filling does not specify an air velocity but requires one to be justified. The specification used is taken from the EU GMP Guide. The specification has been proved effective based on airflow (smoke) study investigations, which have demonstrated that this air speed is of sufficient velocity to remove contamination from Grade A filling machines.

The FDA Guide requires airflow measurements to be taken at 6" from the filter face and "proximal to the work surface". The EU GMP Guide requires readings to be taken at the working height. Working height is defined in local procedures and demonstrated as effective by way of airflow (smoke) studies. The practice at many organisations, during bi-annual re-qualifications, is that airflows will be measured from both locations using the specification detailed in the table above.

In addition to qualifications, many organisations elect to measure airflows periodically before commencing an activity.

Air changes – general guidance

Each clean room grade has a set number of air changes per hour. Air changes are provided in order to dilute any particles present to an acceptable concentration. Any contamination produced in the clean room is theoretically removed within the required time appropriate to the room grade. Monitoring air changes is necessary because the re-circulation of filtered air is important for maintaining control of the clean area.

Air change rates stated are the minimum and should be calculated from supply air volume and room volume measurements. Previous standards have stipulated 20 air-changes per hour as the minimum.

This remains a reasonable set minimum, although frequently more frequent air-changes are desirable.

Clean up times

Connected to air changes is the time taken for a clean area to return to the static condition, appropriate to its grade, in terms of particulates.

Specification	Source of specification	Method
All clean rooms 15 – 20 minutes to achieve static classification	EU GMP / ISO 14644	Particle counting

The conducting of clean-up times is an optional test to be considered at the time of room classification; following substantial changes to room design; for newly built clean rooms or as part of an investigation.

According to EU GMP Guide (1st March 2009 revision) these 'clean-up' times apply to Grade A and B areas only. They may, however, be used as a guidance for Grade C and D areas in the course of an investigation.

Positive Pressure

Connected to the measurement of air flow is positive pressure. In order to maintain air quality in a clean room the pressure of a given room must be greater relative to a room of a lower grade. This is to ensure that air does not pass from "dirtier" adjacent areas into the higher grade clean room. Generally this is 15-20 Pascals, although some areas of the same grade at BPL also have differential pressure requirements due to specific activities.

Specification	Source of specification	Frequency	Method
15-20 Pa relative to lower grade rooms (for A this is relative to B; for A and B this is relative to C; for C this is relative to D)	EU GMP	Monitored every 20 seconds for Grade B rooms; monitored by Production staff for C and D areas	Electronic micro-manometer

Note: Pressure differentials (expressed in Pascals) are the relative pressures from a higher grade area into a lower one. These are guidance values taken from EU GMP Annex 1. There will be some exceptions to the criteria stated above for particular areas which require different pressure differentials.

HEPA filters

HEPA (High Efficiency Particulate Air) filters are designed to control the number of particles entering a clean area by filtration. In Grade A zones HEPA filters also function to straighten the airflow as part of the unidirectional flow. In order to measure the effectiveness of the filters they are checked for leaks. Leakage is assessed by challenging the filters with a particle generating substance and measuring the efficiency of the filter.

Specification	Source of specification	Frequency	Method
99.997% (minimum efficiency)	BS EN 1822	Six monthly	Smoke generator / photometer

Note: HEPA filter integrity taken from BS EN 1822 Parts 1 and 2. The 99.997% efficiency is based on particle sizes 0.3 μm and larger (i.e. theoretically only 3 out of 10,000 particles at 0.3μm size can penetrate the filter).

Temperature, humidity, lighting and room design

Grade B rooms have set requirements for temperature and humidity. These are monitored for operator comfort and to avoid a high temperature – humidity situation which may result in the shedding of micro-organisms. Other clean areas have a temperature appropriate to the process step (e.g. if the process requires a cold room at 2-8°C).

Lighting should be adequate, uniform and anti-glare, to allow operators to perform process tasks effectively. A range of 400 to 750 lux is recommended.

Specification	Source of specification	Frequency	Method
Grade B Temperature: 18 ± 3°C	BPL Site Master File	Monitored by production staff	Thermometer
Humidity: 45 ±15%	ISO14644 -4		Humidity reader

Clean rooms are specially designed rooms. The surfaces are constructed from materials that do not generate particles, and are easy to clean.

ISO 14644 Part 2

With ISO 14644 part 2, there are fewer changes as part of the 2015 revision. The revision of ISO 14644-2 emphasized the need to consider a monitoring strategy in addition to the initial or periodic execution of the classification of a cleanroom or clean zone in accordance with ISO 14644-1:2015.

The main points were:

- Reclassification is now a minimum of annual (this is a change from some areas to be assessed six-monthly). However, it should be noted that EU GMP states aseptic filling to be six-monthly, unless justified.

- Requirement for an on-going monitoring strategy in addition to cleanroom classification. This should be by risk assessment. Those working in GMP facilities should be following this already.

- There is a note that particulate levels are likely to be higher during processing, when compared with classification.

- The tubing length to the particle counter should be less than 1 metre (it was formally 3 metres maximum).

In addition, there is a recommendation that particle counters should meet a standard titled ISO 21501. It appeared, at one stage through the revision process, that this would become mandatory. This standard requires particle counters to have an error rate, at each particle size, of not more than ±20%). Counters assessed against this standard must be certified.

References

A) Standards

- Eudralex. The Rules Governing Medicinal Products in the European Community, Annex 1, published by the European Commission, Brussels: Belgium, 2015
- FDA. Guidance for Industry Sterile Drug Products Produced by Aseptic Processing–Current Good Manufacturing Practice,' *Food and Drug Administration*, Rockville, MD, USA, 2004
- ISO 14644-1 Cleanrooms and associated controlled environments – Part 1: Classification of Air Cleanliness, ISO, Geneva, Switzerland, 2015
- ISO 14644-2 Specifications for testing and monitoring to prove continued compliance by ACP, ISO, Geneva, Switzerland, 2015

B) Other references and further reading

- Akey, J. (2005): 'Info Overload: What Do My Particle Counts Mean?', *Indoor Environment Connections*, Volume 6, Issue 11
- Akey, J. 'Validation, Calibration Assure Investigation Accuracy', *Lighthouse*, USA
- Halls, N. (2004): 'Effects and causes of contamination in sterile manufacturing' in Hall, N. (ed.): 'Microbiological Contamination Control in Pharmaceutical Clean Rooms', CRC Press, Boca Raton, pp1-22
- Johnson, S. M. (2004): 'Microbiological Environmental Monitoring' in Hodges, N. and Hanlon, G. 'Microbiological Standards and Controls', Euromed, London
- Lovegrove-Saville, P. and Perry, M. (2000): 'Setting environmental alert and action limits', PharMIG News, Issue 3, December 2000
- PDA Technical Report No. 13 (Revised): 'Fundamentals of an Environmental Monitoring Program', September – October 1997 (revised 2001)
- Pollen, M. (a) 'Particle Sample Tube Lengths for Pharmaceutical Monitoring', *Lighthouse*, USA

- Pollen, M. (b) 'Airborne Particle Counting for Pharmaceutical Facilities: Update 2008, EU GMP Annex 1', *Lighthouse*, USA
- Sandle, T. (2004): 'General Considerations for the Risk Assessment of Isolators used for Aseptic Processes', <u>Pharmaceutical Manufacturing and Packaging Sourcer</u>, Samedan Ltd, Winter 2004, pp43-47
- Sandle, T. History and development of cleanrooms. In: Sandle, T. and Saghee, M.R. *Cleanroom Management in Pharmaceuticals and Healthcare*, Euromed Communications: Passfield, UK, pp21-38, 2013
- Vincent, D.(2002): 'Validating, Establishing and Maintaining A Routine Environmental Monitoring Program for Cleanroom Environments: Part 1', <u>Journal of Validation Technology</u>, August 2002, Vol. 8, No.4
- Whyte, W. (2001): '<u>Cleanroom Technology: Fundamentals of Design, Tetsing and Operation</u>', Wiley
- Whyte, W. (ed.) (1999): '<u>Cleanroom Design</u>', 2nd Edition, Wiley.
- Wilson, J. (1997): 'Setting alert / action limits for environmental monitoring programs', PDA Journal, Vol.51, No.4, July –August 1997

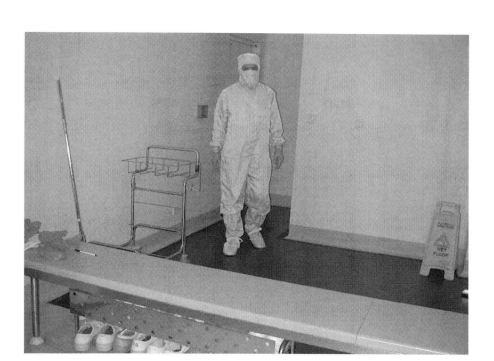

Printed in Great Britain
by Amazon